Ronny the Frenchie Copyright 2023 - All Rights Reserved.
The content contained within this book may not be reproduced, duplicated or transmitted without direct written permission from the author or the publisher.

Under no circumstances will any blame or legal responsibility be held against the publisher, or author, for any damages, reparation, or monetary loss due to the information contained within this book, either directly or indirectly.

Legal Notice:
This book is copyright protected. It is only for personal use. You cannot amend, distribute, sell, use, quote or paraphrase any part, or the content within this book, without the consent of the author or publisher.

Disclaimer Notice:
Please note the information contained within this document is for educational and entertainment purposes only. All effort has been executed to present accurate, up to date, reliable, complete information. No warranties of any kind are declared or implied. Readers acknowledge that the author is not engaged in the rendering of legal, financial, medical or professional advice. The content within this book has been derived from various
sources. Please consult a licensed professional before attempting any techniques outlined in this book.

By reading this document, the reader agrees that under no circumstances is the author responsible for any losses, direct or indirect, that are incurred as a result of the use of the
information contained within this document, including, but not limited to, errors, omissions, or inaccuracies.

TABLE OF CONTENTS

Introduction...4
Leonardo Da Vinci—The Man Of Many Talents!................7
Isaac Newton—Mathematical Genius..............................11
Albert Einstein—Liked To Do Things His Way..................14
Cecilia Payne-Gaposchkin—More Than Stars In Her Eyes.........18
William Perkin—Accidental Fan Of Royals........................20
Wilhelm Conrad Röntgen—Clever And Generous................23
Marie Curie—Multiple Nobel Prize Winner!......................26
George Washington Carver—Changed The Way We Think About Soil...........30
Margaret Knight—She Couldn't Stop Inventing!................33
Mary Anderson—Always Open To Ideas..........................36
Josephine Cochran—Dishwasher Darling!........................39
Stephanie Kwolek—Lifesaver...42
Morton Heilig—A Little Ahead Of His Time......................44
Wilson Greatbatch—Accidental Inventor.........................46
Henrietta Leavitt—Rose Above Expectations For Her Gender...........48
Louis Braille—When He Needed Something, He Created It Himself...........50
Mary Anning—Fabulous Fossil Finder...............................53
Maria Sibylla Merian—Mad About Bugs............................56
Alan Powell Goffe—Vaccine Pioneer.................................59
Asmeret Asefaw Berhe—Helped Feed Her War-Torn Country.........61
Conclusion..63
Freebies..64

INTRODUCTION

Hello kids!

It's Ronny the Frenchie here, your favorite bulldog.

I can't wait to tell you all about some of the most interesting, inventive, clever, kind, tough, hardworking, and persistent people that are so inspiring to me and I'm sure will be for you too!

Each person featured in this book was completely different.

They were all born in different places, to different levels of wealth and opportunity. They all faced problems, whether that was because of their age, background, financial situation, the color of their skin or the fact they were born a female and not a male.

 ### Yet, they all still managed to succeed.

Another thing these super cool people had in common is that, even though they faced many problems on the way to achieving their dreams, even big mistakes and terrible failures, and people telling them they were wasting their time or that they couldn't do something simply because no one else had done it yet, they kept on going.

Why? You'll soon find out!

In fact, you'll find out some of the real secrets to success.

One thing I noticed, and loved is how curious each and every person was, and that's what drove them to their goal in the first place. Curious about bugs, fossils, or technology, or even soil. It was that curiosity that drove them to succeed. I'm such a curious one, too. I just haven't done anything extra special yet.

Oh, hang on, I'm writing this book that kids find interesting and inspiring. So that's a bit special, isn't it? It is!

Okay, enough about me. Get comfortable and enjoy reading about these extra special people. You know what? I think you're probably extra-special too.

LEONARDO DA VINCI
THE MAN OF MANY TALENTS!

I'm sure you've heard of this man;

HE IS ONLY KNOWN AS ONE OF THE MOST GIFTED AND BRILLIANT MEN EVER!

And not just in one area! He was a scientist, inventor, mathematician, sculptor, and engineer, and so many of his ideas in these areas have changed the world.

Leonardo was born in 1452, in Florence. He was also known as being an extremely advanced thinker. It was as if he was centuries ahead of his time. On top of that, he was one of the best painters in the world. Where did this guy find the time?!

Have you ever heard of the paintings: 'The Last Supper' or 'The Mona Lisa'?

WELL, HE PAINTED THEM!

And guess what, he didn't even go to school! He was taught some basic skills like math, writing, and reading at home, but most of his inspiration and skill came from spending time outside, in nature, observing everything around him and trying to figure out how things worked.

SO HOW DID HE GET TO BE SO SMART?

He was just super curious and wanted to learn about everything! He's curious like me! It was when he was 15 that he became obsessed with learning all about the human body. He'd started training with sculptor and painter, Andrea del Verrocchio and found it fascinating.

There's one super interesting thing about Leonardo, though, that you may not know.

HE USED TO PROCRASTINATE, A LOT.

 You know, like he'd put things off and sometimes not even finish things he'd started. Why? Well, he was a perfectionist and so maybe he was overwhelmed with getting something so perfect that it was all too much?

We all know it can be hard to finish projects sometimes, but soon after that, no one would hire Leonardo because they didn't know if he'd finish the job or not! Oopsie!

I mean, for example, once he was hired to sculpt a horse. He actually spent 16 years studying horses in order to get it just right! I hope you don't spend that long on school projects!

You'd think life would be a breeze for someone with so many talents but, just like us, he struggled with day-to-day problems and at one point, after being rejected for a part of his work, felt like a loser. He was so critical of his own work, and despite all he did, he felt he hadn't achieved his full potential.

 We think he did so well; there are only so many hours in a day! Sometimes he failed at things and wondered whether he was any good at all.

CAN YOU IMAGINE?!

And that's because he felt bad for procrastinating and for all those unfinished projects. Some people say he may have suffered from

ADHD (A CONDITION WHERE PEOPLE FIND IT DIFFICULT TO FOCUS AND OFTEN FEEL RESTLESS)

and that could explain a few things, but no one knows for sure as it wasn't tested for back then.

So, we can be inspired by Leonardo's many talents and also maybe learn that it's okay not to be perfect. It's better sometimes to just get the job finished!

ONE TIP TO STOP PROCRASTINATING IS TO BREAK DOWN A BIG TASK INTO LITTLE ONES,

like me taking yummy bite-size pieces of meat rather than trying to eat a huge steak in one go! (Hang on, that sounds good!) Or, you can try starting a project for just 10 minutes. 10 minutes is better than no minutes.

AND USUALLY, ONCE WE GET STARTED, WE KEEP ON GOING.

ISAAC NEWTON
MATHEMATICAL GENIUS

Another amazing mathematician was Isaac Newton. But something you might not know about him is that he was a tiny weak bubba (whose dad died before he was born) and wasn't even expected to survive. And we're so glad he did survive those early days because his discoveries of the law of gravitation truly changed the world we live in!

You'd be totally surprised, like I was, to learn that Isaac wasn't a good student! He started school at 12 (super late compared to today), didn't even like to study and got bad grades. Naughty, naughty.

He was bullied though, and once he fought back. And he won!

 **EVER SINCE THEN, THOUGH, ISAAC WENT FROM HAVING SUPER LOW GRADES TO TOPPING THE CLASS. INTERESTING, HEY?**

When Isaac was 16, his mother took him out of school and he was made to work as a farmer, which he hated. But the good thing is, he was so, so bad at farming, that he was eventually sent back to school and went on to study at the famous Cambridge University.

To think his brain could have been wasted growing potatoes?! You know what though, his mum would not pay for his university fees, so he had to work as a servant while studying at the same time; **NOT EASY!**

What a rollercoaster time for Isaac. He got the good luck of receiving a scholarship, but then the bad luck of an outbreak of a virus, and the university, unfortunately, closed down for two years. He didn't waste his time though, and spent a lot of time thinking and observing and making discoveries in physics, math, and astronomy.

He didn't waste his time though, and spent a lot of time thinking and observing and making discoveries in physics, math, and astronomy. Having been in the countryside, he'd have had a great view of the night stars! And it was during this time that he had his fabulous experience of sitting under a tree and watching an apple fall, which inspired his law of universal gravitation.

 HE PUBLISHED A FAMOUS BOOK CALLED 'MATHEMATICAL PRINCIPLES OF NATURAL PHILOSOPHY.'

It made him become famous and he was even knighted by the Queen! Very fancy and important and, from then on, he was called Sir Issac Newton.

DID YOU THINK THAT ALL FAMOUS PEOPLE WERE OUTGOING AND POPULAR?

That isn't true at all. Newton was a shy one, and very clever and curious. I see a trend here—curious and smart! Maybe I'm a genius?!

I THINK, YOU SHOULD BE WHO YOU ARE AND LET YOUR CURIOSITY LEAD YOU TO WHERE YOU BELONG!

ALBERT EINSTEIN
LIKED TO DO THINGS HIS WAY

When little Albert was born, his parents were worried, very worried. His head was a little big and misshapen, which led his parents to think something was wrong with him. But, it became normal after a few weeks. Phew! Now, you would think that Albert, being a genius and all, would have been way ahead of all the other kids his age.

NOPE! HE WAS SLOW TO LEARN TO TALK, AND AGAIN, HIS PARENTS WERE WORRIED ABOUT HIM.

He was slow to speak and to talk to those around him, and eventually spoke at the age of four.

He'd practice his sentences in his head first till he got them right before speaking them. Let's just say, his parents did not have high hopes for Albert.

A turning point in his life was when his dad gave him a small compass. Maybe in those days, it was a cool new gadget, and Einstein loved it SO much. It's what sparked his love of science, which was more of an obsession. Not long after that he began playing the violin, another love of his that he kept up all through his life.

You might think Einstein was not good at school like the two clever ones you read about above, but he was! Very good,  and almost top of the class—in math and science. He loved those subjects so much that he started teaching himself and by 12 he taught himself Euclidean geometry.

SOUNDS TRICKY! HE ALSO MASTERED CALCULUS (ALSO PRETTY TRICKY).

Albert preferred to teach himself though, and not be taught in the strict style of the teachers back then. It annoyed him so much, that he ended up leaving school and went away from Germany so he didn't have to join the army like they made all the young boys do back then.

$ax^2 + by^2 + c = 0$

I don't blame you for that, Albert! This made his poor parents worry again because they knew he was clever, and wondered what he would do next? He didn't want to stop learning and applied for a place at the Swiss Federal Institute of Technology, but instead of acing the test as you'd imagine, he failed!

YEP! EINSTEIN FAILED. THOSE TWO WORDS DON'T LOOK RIGHT TOGETHER, HEY?

I'm seeing another pattern here; curious, clever, some failure and some success. Failure is right up there! And totally normal guys. But there was a reason why he failed. He excelled in the subjects he loved and passed math, but hadn't studied botany, zoology and language; those subjects he did not enjoy as much.

Goes to show, we gotta do some stuff we don't love in order to get to do the stuff we do love. Right? Like how I need to shake hands with my owner in order to get a treat. Well, not quite like that but similar! Einstein was better prepared the next time and got accepted. So once he finished studying you'd think he'd be snapped up and get a job quickly, but he took a few years to get a job, as a patent clerk in 1902. He wasn't offered the position of professor until 1909. Why did it take so long for someone so talented?

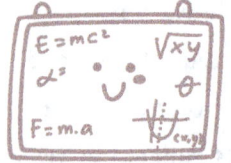 Well, people didn't recommend him for jobs because he was a bit disruptive in the classroom, often asking the professors tricky questions and arguing too. Let's just say the professors were not too impressed with this out-of-the-ordinary behavior, which was a problem because they were the ones who had to write nice letters of recommendation about him. I think he liked to do things his way. Which is great, but sometimes we have to follow some rules in order to do what we really want to do. It's life, huh?

Einstein decided to get a job that was boring but easy and save all his brainpower to research and study in his spare time. This was when he developed his theory of relativity.

AND HIS FAMOUS $E = MC^2$.

He was even awarded the Nobel Prize in Physics in 1921 for his clever explanation of the photoelectric effect. Yep, all the while working at a boring old job! And how proud must his parents have been? And a lot less worried! Albert followed his own path and it led him right to success. He also wrote lots of books and knew a lot about philosophy too.

 **THERE'S ALWAYS SOMETHING TO LEARN, HEY?**

CECILIA PAYNE-GAPOSCHKIN
MORE THAN STARS IN HER EYES

This famous lady from England loved science too. And the area of science she loved the most was space. She wanted to become an astronomer. But can you believe that back then, when Cecilia finished her studies in 1923 (at Cambridge University), she did not get a degree because Cambridge did not give them to women at that time? She was given a certificate instead. Not quite the same. Hmmph! I bet they learned their lesson when **CECILIA ACTUALLY DISCOVERED WHAT THE SUN IS MADE OF! HOW COOL IS THAT?!**

Back then, people thought it was made of similar things to our Earth, just that those materials in the sun were hotter. Stuff like iron, etc.

Well, that would make sense because there's so much iron on Earth. She worked out that the sun and stars are actually made of oxygen and hydrogen and helium by the way!

BUT CECILIA WOULDN'T GIVE UP, NO WAY!

She moved to America to Radcliffe College, which was a university for women only. It only took her two years to prove everyone wrong about what the stars and the sun are made of. You see, the sun is 93 million miles away (I know!!) It is so hot that we couldn't get near it to investigate. According to Nasa, its core is 27 million °F! And the surface is around 10,000 °F — still scorching! So how did Cecilia manage to work this out? Well, she figured it all out by looking at the color of their light. Interesting, hey?

Unfortunately, it took years for her to have people believe her ideas; mostly because she was a woman and it wasn't common for women to be such amazing astronomers. She eventually became the first female professor at Harvard University. I think it was so great that she didn't give up on becoming what she wanted to be because of something like whether she was a male or female.

NOTHING CAN STOP US FROM GRABBING OUR DREAMS. NOTHING, GUYS!

WILLIAM PERKIN
ACCIDENTAL FAN OF ROYALS

While some people spend hours and hours or years and years working on their theories and figuring out how things work, some things are discovered by accident.

A LUCKY ACCIDENT!

One such lucky accident happened way back in 1856, all because of some black sticky and stinky leftover coal tar that people threw away because it was useless.

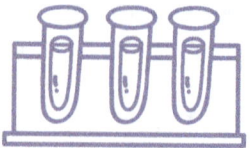

Back then a man named Wilhelm von Hofmann realised there was some potential in tar to be used in medicines, and tasked the young William, his assistant, to do some experiments in making quinine (used to treat malaria) more cheaply. He experimented with coal tar—because it has similar properties in it to quinine, made from more expensive substances, like the bark of exotic trees.

OF COURSE, HE FAILED AT FIRST. THAT'S HOW IT GOES, RIGHT?

One time, when he was mucking about mixing this and that, the mixture went a bright purple.

HE LIKED IT!

He realized it could be used to dye delicate silk and thought he could make some money with this idea! You might think, 'but didn't they have purple dyes back then?' Well, they did, but it was made from crushing a specific type of exotic snails! Only the rich people could afford that.

But now, William could make purple dye from something that was thrown out, that was free, and everywhere. He called the color mauve, started his own factory and thought he'd make some money. He had two big fans of his work too!

QUEEN VICTORIA AND NAPOLEON'S WIFE BOTH LOOOOVED THE LOVELY MAUVE COLOR.

Because it was then not crazy expensive, lots of people could then wear it and (you guessed it) the color became super popular and fashionable.

WILLIAM BECAME SUPER-RICH, BY THE WAY, AND NEVER HAD TO WORK AGAIN!

So, I wonder if you might make an awesome discovery by accident?! The fact is, William kept on trying till he came up with something, and we can definitely learn something from that. And even if we don't achieve exactly what we set out to do,

IF WE KEEP OUR EYES OPEN FOR OPPORTUNITIES, WE COULD STUMBLE ON SOMETHING WONDERFUL.

WILHELM CONRAD RÖNTGEN
CLEVER AND GENEROUS

I'm pretty sure everyone is impressed when they look at an X-ray. I mean, a photo of inside the body, that we rarely get to see. It's still impressive now, but back when it was invented, it must have really wowed everyone.

WHAT YOU MIGHT NOT KNOW, AND THAT I JUST LOVE, IS THAT THIS IS ANOTHER ACCIDENTAL INVENTION!

Back in the early 1890s, scientists were busy researching the cathode ray tube—(light emitting tubes that used to be used in TVs and computers). One day, Wilhelm was in his lab experimenting with cathode ray tubes.

AS YOU DO!

Anyway, he figured out that barium platinocyanide glowed, even though it was covered in a thick layer of cardboard. This got him thinking because, yep, he was another curious scientist and just had to figure out why it still glowed and was not dimmed by the cardboard. He experimented and tried and failed and realized it must come from some kind of radiation. Some kind of rays must be travelling about to make this happen.

HE CALLED THEM X-RAYS. (THE X STANDS FOR 'UNKNOWN UNTIL I FIGURE IT OUT!')

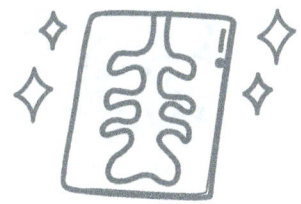

He made an X-ray machine and wanted to test it out. And who better to test it out on than his wife?! The first X-ray ever was this X-ray on her hand. It worked, and his discovery became so valuable in the medical world. It could be used to be able to see exactly where a break of a bone was, for example.

◇ **WILHELM'S DISCOVERY WAS SO USEFUL IN FACT THAT HE WON THE NOBEL PRIZE FOR PHYSICS.** ◇

And you know what? What I love most about this guy, is that he didn't even get a patent on his invention (wanting the rights to the invention and gaining money from it).

☆ **WHAT A GENEROUS PERSON!** ♡

I really like his curious mind and his showing us that being generous is just as cool as inventing something amazing.

MARIE CURIE
MULTIPLE NOBEL PRIZE WINNER!

Meet Marie Curie! She was the very first woman to win the Nobel Prize.

I BET SHE FELT PRETTY SPECIAL, AND INDEED SHE SHOULD, BECAUSE NOT ONLY DID SHE WIN ONE NOBEL PRIZE, BUT TWO!

And, they were in different fields.

Marie was born in Poland in 1867 and she grew up to be a very skilled physicist. It helped that her parents believed that education was very important, and also that her dad was a math and physics teacher.

He sure influenced the things Marie became interested in and she did so well at school that she finished early, at 15. The thing is, she wanted to study more but, just like Cecilia Payne, women were not allowed to go to university in Poland.

GRR! I DON'T THINK THAT'S FAIR AND IT MAKES ME VERY MAD!

But here's the bit you might not know that I think is so cool! She attended a 'flying university' which was like a secret club where their members meet and study and talk about all the interesting things like politics and things going on in the world. So that they didn't get caught, they moved around to different places at night.

I BET THAT WAS KINDA TOUGH, HEY?

I forgot to mention, Marie's sister loved studying and went to the secret meetings too and finally, they were both accepted to go to university in Sorbonne, in France. The thing is, they didn't have enough money to go at the same time, so they had to take turns! When her sister Bronya was studying, Marie worked as a governess and tutor to earn money for her sister's studies. She couldn't resist studying though and used any spare time to study chemistry and physics.

Can you believe that Marie had to wait six years till her sister finished her studies? But, the wait was worth it because her sister had become a doctor and was happy to pay for Marie's studies.

IT WAS HER TURN! SHE WORKED SO HARD.

She sometimes even forgot to eat and didn't sleep much at all, which I think is, erm, well, that can't be good for you. I need my sleep! And my food! But she got her degree in physics and was top in her class. Of course, she was!

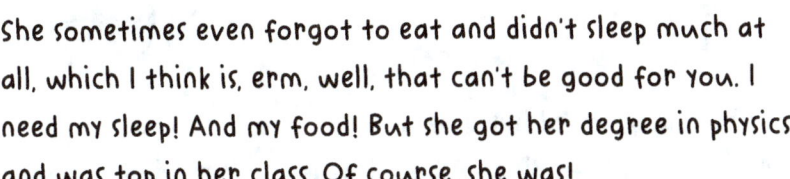

THE NEXT YEAR SHE WHIZZED THROUGH AND GOT A DEGREE IN MATH. SUPER FAST!

Would you like to hear how she got her surname? No, she wasn't born with it; it's her husband's name! Yep, she needed a laboratory to do all her experiments in, and someone introduced her to Pierre Curie who let her share his. He was a physicist too.

AND THE REST IS HISTORY!

Marie continued to study and expanded on research already done by Becquerel who discovered that rocks that had uranium in them gave off a certain light (or rays).

'WHY DID THIS OCCUR?' WONDERED MARIE.

 She was determined to find out, and eventually found out that it was changes in the atoms in the rocks that made them give off light. No one knew that atoms could change in this way.

MARIE CALLED IT 'RADIOACTIVITY'.

Her husband helped her research too and she published all of her amazing findings. All of that led to her winning the Nobel Prize. She shared this prize with Becquerel and her husband.

 ### THAT WAS GOOD OF HER!

Sadly, her husband died in an accident, but she kept on with her work and eventually was given her husband's job, and became the first female professor at Sorbonne. After that, she received another Nobel Prize, this time for chemistry, and expanded ways to use X-rays to help injured soldiers in the war. She did so many good things and we can see that if you are determined and patient and believe in your work,

YOU CAN DO AMAZING THINGS.

GEORGE WASHINGTON CARVER CHANGED THE WAY WE THINK ABOUT SOIL

Back in the late 1800s in America, it was a hard time for people like George, because he was a black man, and slavery was still happening. (Slavery was when people owned other people).

GRR... I GET SO MAD! I NEED TO STOMP MY LITTLE PAWS!

Did you know that black people were not allowed in some parts of the country, or go to school and many were forced to be farmers, without being paid?

Thankfully slavery eventually ended, but black people still had a really tough and unfair time and George had to travel all around just to be able to get an education. He'd always loved books and reading though and loved plants and music, and would not let the ridiculous rules regarding black people get in his way. He first studied farming and wanted to be a teacher. He was interested in the nutrients in the soil. While not being the first one to discover the process of crop rotation, (growing cotton was taking the nutrients out of the soil and over time, there wasn't much nitrogen left and nitrogen is super important for growing plants),

CARVER MADE BIG LEAPS IN THIS IDEA, AND TOOK IT TO ANOTHER LEVEL.

He deeply researched it and promoted the idea and process so that more people could benefit from rotational crops to improve the health of the soil.

He specifically showed that if you grow peanuts in the same soil for a while, the peanuts put some nitrogen back in the soil (they have some kind of special bacteria on them). With his promotion of this work, people started growing lots of peanuts and sweet potatoes (which do a similar thing).

There were a LOT of peanuts and sweet potatoes after this discovery of his, and so he figured out how to use them to make heaps of things like shampoos, body creams and even medicine.

 **IN FACT, HE DISCOVERED OVER 300 USES FOR PEANUTS!**

I wonder why he didn't think to invent peanut butter?! Well, he greatly changed how people farmed and used soil and he had a big impact on agricultural practices in the Southern United States. I love that George Carver did not let the bad ideas of some people stop him from going after his love of learning.

IF YOU WANT TO DO SOMETHING DEEP IN YOUR HEART, DON'T LET ANYTHING STOP YOU!

MARGARET KNIGHT
SHE COULDN'T STOP INVENTING!

Back in around 1868, Margaret used to work in a shop that made paper bags. But Margaret thought the design could be greatly improved. And she had a great idea. So, she invented a machine that made a flat-bottomed bag. Good on her! Now, you might think, what's so special about flat-bottomed bags?

WELL, THEY CHANGED EVERYTHING!

You could easily place them on a table for easy unpacking and they didn't break, being much stronger. All the major shops adopted them.

Plastic bags took over for a while, but we're seeing flat-bottomed bags more and more these days, as gift bags, carry bags, and McDonalds still uses them!

✧ EVEN AS A KID, MARGARET WAS A NATURAL INVENTOR. ✧

Once, when she was 12, she saw someone get hurt in a factory. So, she invented a stop-motion device that would help protect people in the future from getting injured in this way, and this item was used in many factories after that!

⚡ SHE MUST HAVE FELT SO GREAT TO HAVE HELPED SAVE PEOPLE FROM TERRIBLE ACCIDENTS IN FACTORIES. ⚡

Margaret was a natural inventor and she invented many other things like a shoe-making machine, a numbering device, and more.

Oh, get this. One person said that it was them that invented the flat-bottomed bag machine that she had invented. They tried to get away with this by saying that women couldn't possibly create something so complex.

 **CAN YOU BELIEVE THAT?! GRRRRR.**

Anyway, Margaret had the sketches of her idea to prove that she did indeed invent it, and she won the law case.

Even when Margaret was 75 years old she was apparently working on her 89th invention! Let's just say, she couldn't stop! Her imagination and creative brain were always buzzing. But what a wonderful skill. I wonder guys, if you have any ideas for inventions?

IF YOU DO, WHY NOT GIVE IT A GO AND TRY TO CREATE THEM?!

MARY ANDERSON
ALWAYS OPEN TO IDEAS

CAN YOU IMAGINE CARS WITHOUT WINDSCREEN WIPERS?

How could people possibly drive without them if it was raining? I have no idea, but thankfully businesswoman, Mary Anderson had the brains to invent them! There weren't too many businesswomen back then either, so I think she's extra special!

One day, Mary was riding something similar to a streetcar in New York when she noticed that every now and then the driver had to get out of the streetcar and clean the snow off the windows so he could see.

I still can't believe no one thought of this earlier but, anyway, let's carry on!

So, Mary must have thought it was a bit silly to have to keep stopping all the time and not have wipers on the tram itself. It would also make the journey faster without all of that stopping. I'm sure the tram drivers would prefer not to have to jump out into the snow all the time, too. So, when she got home she sketched a design for wipers. It had a little lever inside the tram that the driver could pull to lift the wipers. Handy!

She soon got a patent for her idea (so no one else could steal this idea). But when she tried to get the wipers made at manufacturers they said they wouldn't work, or sell. Mary's great-great niece thinks this happened simply because she was a woman.

 **GEE, WOMEN HAD IT TOUGH IN THE OLD DAYS. WHAT AM I SAYING?**

They still have it tough in many ways. But I'm glad things have improved!

Moving along... so, around 20 years later when cars were being made in huge numbers,

THEY USED MARY'S DESIGN!

Because her patent had expired, they got away with it, saying it was their idea, and Mary didn't get any money for her super-useful invention.

 **HMMPH!**

Anyway, being the clever woman that she was, she was successful anyways as a rancher and property developer. I guess we can think of Mary when we have ideas, to just trust them no matter what. And also, to not put all your eggs in one basket. In other words, if one idea doesn't work out, keep on using your brain and skills and do something else in the meantime. Don't bet all your eggs (money and time) on one thing. And hey, even though she wasn't credited for her invention nor did she get any money from it,

HER INVENTION WAS SAVING LIVES EVERY DAY, AND THAT'S WORTH MORE THAN MONEY, RIGHT?

JOSEPHINE COCHRAN
DISHWASHER DARLING!

Back during the industrial revolution when all sorts of machines and things were being invented to make life easier, and totally changing the way people lived, women were still having to do a lot of their work by hand. Where were the machines to help with domestic chores that never ended?

Well, one was on its way, thanks to Josephine Cochran. She didn't have to struggle with the hard work of washing, coming from a wealthy family, but she still saw how hard it was.

Instead of washing dishes or clothes all day, she loved to entertain.

 ## LUCKY FOR SOME!

She had maids to wash her dishes for her. Josephine loved her beautiful precious china plates and cups (that were hundreds of years old), which to her anger, often got chipped when washed by the staff who worked for her. She thought they must have been so rough with their washing technique. Josephine loved this china so much, in fact, that she decided to wash them herself, just to be sure they didn't get damaged.

 ## CAN YOU GUESS WHAT HAPPENED? SHE GOT TIRED OF WASHING ALL OF THE DISHES!

There needed to be something that could clean them more gently. Back then, the wealthier people that surrounded Josephine didn't really need a dishwasher though. Many people had maids to do all that work (I'm sure the lower class would have loved a dishwasher!) and when Josephine talked to her husband about the need for a machine, he didn't think dishwashing was of any importance.

So, Josephine took his advice and forgot about her dishwasher idea. Just kidding! She set right to work designing a machine.

YEP! SHE SPENT MONTHS AND MONTHS ON IT.

When the design was ready, she got a mechanic to help her put a model together.

 **IT WORKED!**

It wasn't easy to sell though at first but, eventually, at the Chicago World Fair, she found lots of buyers from restaurants and hotels. And now, we even have dishwashing machines in our homes, thanks to Josephine and her determination to make washing dishes easier and better. The thing we can really learn from her is that, while some people might think our idea is not important,

WE MIGHT REALLY BE ONTO SOMETHING AND WE SHOULD TRY IT OUT ANYWAY.

STEPHANIE KWOLEK
LIFESAVER

Young Stephanie loved science and clothes. Hmm, nice combination. I wonder what becomes of someone like that?

WELL, YOU JUST WAIT AND SEE!

Stephanie set out to get a chemistry degree first, and she wanted to become a doctor. Unfortunately, she didn't have enough money for the degree.

NOT FAIR, HEY?

But she went and worked at an interesting company (DuPont) that made synthetic fibers and materials. It was here that Stephanie experimented with making different types of fibers. She had some ideas, but they were a bit different to what other people were doing at the time, and the people she worked with were worried that her material might clog up their machines. But she trusted her gut feelings, and ended up making a fiber that was so, so strong, (five times stronger than steel!)

AND IT WAS EVENTUALLY NAMED KEVLAR.

It was an awesome product, and was used in so many things, such as bulletproof vests (yep, I said it was strong!) It's kind of cool to think that right now we might think we'll grow up to be a doctor or a teacher or whatever it is,

AND THEN DO SOMETHING COMPLETELY DIFFERENT!

And, it's turned out that since the 1980s, Stephanie's invention has saved so, so many lives with these vests: used by policemen, soldiers and journalists in war zones, and more.

WE CAN SEE HERE THAT WE CAN ALWAYS KEEP OUR EYES OPEN FOR DIFFERENT OPPORTUNITIES THAT WE MIGHT NOT EXPECT.

MORTON HEILIG
A LITTLE AHEAD OF HIS TIME

In 1957, Morton Heiling had the idea of creating a special cinema experience, and so he invented the Sensorama.

BUT WHAT IS IT?

It was pretty ahead of its time, that's for sure! It was a 3D video machine where people could really feel that they were immersed in the experience of, say, riding a motorbike with 3D vibes.

The experience included a wind machine in order to make 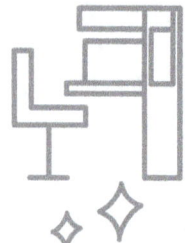 people feel that they were really riding a bike with the wind in their hair. He even gave audiences the experience of watching a belly dancer and sprayed perfume into the room to make the experience come to life!

I THINK IT'S A GREAT IDEA, LIKE OUR 4D CINEMAS TODAY!

But back then, no one wanted to buy this idea. Poor Morton. He was in the wrong era! But what I love about Morton is that he didn't give up and instead invented a 3D headset (not unlike the Oculus that is super popular today) but unfortunately, it seemed the world was not ready for that either, and Morton actually died before seeing his vision (literally) come to life. Sometimes though, it's all about timing, and ideas really need to match what is needed at that time. Morton did pave the way though for future inventions which is still a real achievement. Every little bit of inventive thought is never wasted, and if you put your ideas out there in the world,

YOU NEVER KNOW WHAT MAY COME OF THEM NOW, OR LATER!

WILSON GREATBATCH
ACCIDENTAL INVENTOR

IT'S INTERESTING HOW WE ARE ALL CURIOUS ABOUT DIFFERENT THINGS, DON'T YOU THINK?

Wilson, for example, was interested in listening closely to the human heart. It was very tricky to do that and he didn't have much luck with the device he made at first. Instead of being able to hear the electrical pulses of the heart, the device gave off pulses.

THAT'S NOT WHAT HE WAS AIMING FOR!

But here's another example of trying to invent one thing and inventing something else instead. That is, only if you're able to look at what you've created in an open-minded way and see what opportunity exists for that thing. Or, in other words, instead of scrapping what you've created because it didn't match your original goal, you can think about how it can possibly be used for something else. And that's exactly what Wilson did. His invention turned out to be a pacemaker, which is an awesome creation that helps so many people who have problems with their hearts. Wilson has helped save so many lives as a result of his accidental invention that thankfully he didn't see it as a failure!

OH, I FORGOT TO MENTION SOMETHING SUPER IMPORTANT!

He initially tested his pacemaker out on dogs. I just can't think of a better animal to test out a handy—I mean, life-saving—gadget!

HENRIETTA LEAVITT
RoSE ABOVE EXPECTATIONS FOR HER GENDER

Back in the late 1800s, Henrietta attended Radcliffe College (connected to Harvard University).

YES, WOMEN COULD GO TO UNIVERSITY BY THEN, YAY.

She also worked at the Harvard College Observatory. At work, Henrietta joined a group of other women whose job was to sort out and classify stars using images. Unfortunately though, as women, they were not allowed to use the microscopes themselves and received a really low rate of pay for their work.

I FEEL A HUGE GRRRRR COMING ON!

Anyway, while she was happily studying stars she noticed some patterns that occurred and this discovery resulted in 'Leavitt's Law.' This law explains that a star that takes longer to pulse, is intrinsically brighter than a star that pulses quickly.

NOW WHY IS THIS SO AMAZING?

Well, this discovery meant that astronomers were then able to figure out how far away that star is by the rate of its pulsing. This was a huge thing in the world of astronomy. It led Edwin Hubble in the 1920s to work out that smudges of light in the sky were actually entire galaxies, and this meant that the universe was much, much bigger than anyone realized. Again, one discovery can lead to more discoveries—discoveries that change how we think about the world. Simply by being a woman, Henrietta was limited in what she could do when it came to her love of astronomy. And yet she kept her eyes open for opportunities and still figured out something truly amazing.

THERE CAN BE DISCOVERIES AND INVENTIONS IN EVEN THE SO-CALLED 'LOWER LEVEL' JOBS IF THE MIND STAYS OPEN TO THE POSSIBILITIES.

LOUIS BRAILLE
WHEN HE NEEDED SOMETHING, HE CREATED IT HIMSELF

Little Louis grew up in France in the early 1800s. Yep, a long time ago. When he was three years old, this little boy lost the sight in one eye after playing with an awl (a pointed tool).

POOR LITTLE GUY!

By the age of five, he was blind in both eyes because of an infection that occurred from the first accident. He was a determined little thing though and didn't let his blindness get in the way at all and tried hard at everything he did.

This great attitude led him to get a scholarship to attend a college for blind youth. Back then, blind people learned to read by feeling raised letters on a page with their fingertips.

 **MAKES SENSE!**

But Louis found this style of reading much too slow. Instead of waiting around for someone to come up with a better, faster system, he created his own! By the time Louis was 15, the first braille method was finished. It was a method using raised dots and dashes, similar to what soldiers could read secretly in the dark at night. I must add here, there were SO many cool inventions due to the war! Braille's method was published in 1829.

 WHAT AN ACHIEVEMENT!

To think how he helped so many other blind people to be able to read faster and learn more.

What I think is so very interesting is that Braille made the dots using an awl, the same thing that he damaged his eye with!

 SOMETHING HORRIBLE TURNED OUT TO BE A WONDERFUL TOOL.

He also used a flat slate to poke the dots through to keep them all straight. Braille even created a system for reading music.

 BECAUSE HE LOVED MUSIC, THAT'S WHY! I LOVE THIS GUY.

He showed us that if we want something done, we can just go and do it. A bit of a sad thing is that Braille died before his work became widely used so he didn't get to see the success of all his hard work, but he must have thought it was a game changer for blind people such as himself. And it certainly was.

MARY ANNING
FABULOUS FOSSIL FINDER

Mary was born in 1799 to a poor family on the shores of England. One of her favorite things to do was meander around the beach with her dad, who loved to search for fossils from which he could earn money when he sold them to tourists.

A horrible thing happened though. One night Mary's dad slipped and fell. Tuberculosis got him not long after that and he didn't survive.

The family were not only terribly sad, but they had no income as her father could not work. It was a double tragedy. To help out, Mary kept on selling fossils, and one day while out with her brother, she spotted an unusual-looking skull in the cliffs. It wasn't a typical animal skull.

 ## WHAT COULD IT BE?

Mary was 12 at the time and spent ages trying to find the rest of the bones and digging them out. Thanks to her dad's teachings, she knew what to do. It turns out she hadn't found a little fossil,

SHE'D FOUND A SKELETON OF A DINOSAUR!

A reptile called the Ichthyosaur. And it was an extra cool one. It was half lizard and half fish! A really interesting addition to this story is that not only had Mary found an amazing dinosaur that was over 90 million years old, but she also had geologists coming to her for advice about her discovery. Around 10 years later, Mary found another dinosaur, but people thought it was a fake! Thankfully, she was proven right, but what I can't believe is that because she was a woman and there were not many (if any) female geologists around at the time, she wasn't given any credit for her discovery.

When people talked about it, they didn't even mention her name! But like all the other females we've talked about so far, she did not give up and kept on searching for those beautiful ancient fossils.

SHE LOVED THEM SO MUCH THAT SHE EVEN STUDIED FOSSILIZED POO!

Despite all her efforts, things didn't turn out so well. Mary was not widely recognized for her contribution until later in life. During her final days, she was finally recognized by the Geological Society of London who made her an honorary member for all of her achievements. Her work was written down for all to see. And you know the phrase, 'She sells seashells by the seashore?'

WELL, SOME PEOPLE SAY THIS WAS ABOUT MARY ANNING!

This lady didn't give up on her dreams of searching for dinosaurs when she was not recognized or respected. Like Mary, we should also never give up on our dreams, no matter what anyone thinks because, who knows, you could give a wonderful gift to the world, just like Mary gave a gift to the world of paleontology (study of ancient life).

MARIA SIBYLLA MERIAN
MAD ABOUT BUGS

I think bugs are pretty cute, don't you? And I love to chase butterflies. They're so pretty! But around 300 years ago, insects were called the 'beasts of the devil!'

And no one knew how insects reproduced and thought they just appeared out of rotting meat. Gosh! Now let's move on to something much nicer—Maria. She was one of the world's first ecologists. If that's not enough, Sir David Attenborough totally adores her and thinks she's amazing for what she's taught us about entomology—the study of bugs.

Maria was born in Frankfurt in 1647. Her parents were artists and so it's no wonder she loved drawing.

BUT GUESS WHAT SHE LOVED TO DRAW? BUGS, YEP!

After she got married and had two little girls, it was just expected that she would become a housewife. But Maria went against what everyone said she should be doing as a woman and published a stunning book on flowers. She drew pictures that were full of things people had no idea about. For example, she drew how a moth went from an egg to a caterpillar, and how it flies out of the cocoon it's been growing in (not out of rotting meat) and into the sky!

SO INSECTS WEREN'T BORN FROM DEVILS AFTER ALL AND WHAT A DISCOVERY TO BE ABLE TO SHARE WITH PEOPLE!

Maria didn't stop there. She was known to be the first woman to travel in the name of science and went by ship with her daughter all the way to exotic South America where they went right into the jungle!

They put up with the bites and itches because, for Maria and her daughter, this was heaven for nature lovers. It was teeming with unusual bugs, and plants and flowers.

This adventure led her to publish a famous book on bugs from that area,

'THE METAMORPHOSIS OF THE INSECTS OF SURINAME.'

It was jam-packed with stunning and very detailed pictures of the different stages of the bug's journey from egg to bug and on or around the actual plants they lived on. This was super important because it showed that all buggies need particular plants around them to grow and thrive. They don't just appear and die! Her work is still relevant today, 300 years later, as scientists use her discoveries to learn about the species and adapt or struggle with climate change. Maria combined her skills in drawing with her curiosity and love of bugs.

 ### AND THE RESULT WAS WONDERFUL.

I wonder whether, if you have a skill you could combine it with something you're curious about—like if I had the skill of drawing, I'd make a book all about the different types of dogs other than Frenchies because I'm so curious about other dogs!

ALAN POWELL GOFFE
VACCINE PIONEER

Little Alan was born in England to a Jamaican dad and an English mum. Like many kiddies, he was inspired by his parents who were both physicians. There are so many things you can study in medicine, but he decided to focus on pathology—and learn all about diseases.

AND WE'RE SO GLAD HE DID!

Around that time, the disease, measles, a contagious virus, was really serious. Many children (and adults) were sadly dying from it.

But Alan worked hard on creating a vaccine to help give kids immunity to this terrible disease. When someone is vaccinated they are given a milder dose of a disease. The body's immune system immediately goes about fighting this disease and as they do so, they are teaching the antibodies (the fighter cells) how to overcome this infection. And so now that the child has been vaccinated, if they do catch measles, their body already knows how to fight it and will do so quickly—the child will not get seriously unwell like they would have done when the antibodies didn't know how to fight it.

ALAN WORKED ON A VACCINE FOR POLIO AS WELL.

He tested his vaccine extensively, even testing it out on himself and his family so people felt confident that it was, indeed safe. To think of the many, many lives he saved by developing this vaccine, not to mention paving the way for creating vaccines to prevent future diseases.

 I LOVE THAT IF WE JUST FOLLOW OUR GOALS AND LEARN AND DREAM, WE CAN DO AMAZING THINGS TO HELP OUR WORLD.

ASMERET ASEFAW BERHE
HELPED FEED HER WAR-TORN COUNTRY

Asmeret grew up in Africa, in Eritrea at a time when a war was going on around her. Despite all of that going on, Asmeret was one of only one thousand students to be able to go to university in her country. Now, you've already seen how different people are interested in different things, from stars to fossils to bugs. Asmeret was interested in—soil.

NOW YOU MIGHT THINK, HOW CAN YOU BE INTERESTED IN DIRT?

I think it's fun to bury bones in dirt and dig dirt up. When Asmeret was growing up, there were landmines (small bombs) going off in her country. She wanted to find out what effect these had on the soil, as soil is super important.

WE GROW OUR FOOD IN THE SOIL!

These landmines leaked toxic chemicals into the soil and changed its structure. Not good! I wouldn't want to dig up a bone from around there! Her study led her to find out that if they could just remove one-quarter of the landmines, it would improve the soil so much. It would also allow enough food to be grown in that area to feed 1.6 million people each year.

THIS KNOWLEDGE WOULD HELP SO MANY PEOPLE!

Asmeret grew up in a difficult place, but she was still able to find something good that she could do to help the situation. There's always good we can do in a bad situation if only we look for the opportunities.

THERE ARE ALWAYS PEOPLE TO HELP, TOO.

CONCLUSION

Okay, guys! I hope you had fun reading all about these really inspiring people.

I'm feeling so inspired now! I feel like inventing a delicious bone that buries itself and never runs out of flavor! Are you feeling inspired too? I wonder who you enjoyed reading about the most? I can't decide. They were all interesting and clever and wonderful in different ways.

What I found so amazing was how hard they worked at what they wanted to achieve and how they wouldn't let anyone get in their way!

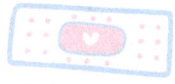

Dear Parents,

Did your children enjoy the inspiring stories about scientists and inventors? Wait, there's more!

Join me once again and dive into the captivating stories of extraordinary sport heroes and fearless entrepreneurs. I can't wait to share their remarkable tales of innovation and determination with you. In addition to the inspiring stories, I have included some fantastic coloring pages that will spark your children's creativity too!

So, what are you waiting for? Claim the freebies by scanning the QR code on the next page or type riccagarden.com/ronny_freebies into your web browser.

Note: You must be 16 years or older to sign up, so grab your parent for help if you need to.

Your Frenchie,
RONNY

CHECK OUT MY OTHER BOOKS

Fun and Amazing Facts for Kids

Set off on a journey of discovery with Ronny the Frenchie, everyone's favorite canine friend—he's got an insane sense of humor and his nose to the ground, sniffing out the most interesting trivia.

Did you know a frog drinks water through its skin—even though it's got that giant mouth? Did you know that some lakes can actually explode? Your kids will not put this book down once they start reading—that's a guarantee!

The Lettering Workbook fo Kids

Join Ronny on a journey to learn the skill of calligraphy and hand lettering: artwork that is always admired, appreciated, and never goes out of style.

This step-by-step guide is easy to follow, filled to the brim with multiple lettering styles, tips and ideas, and is fun all the way through!

A long list of tools is not required, just a wish to get creative and put in some practice. This will bring a reward of relaxation and creating some unique artwork.

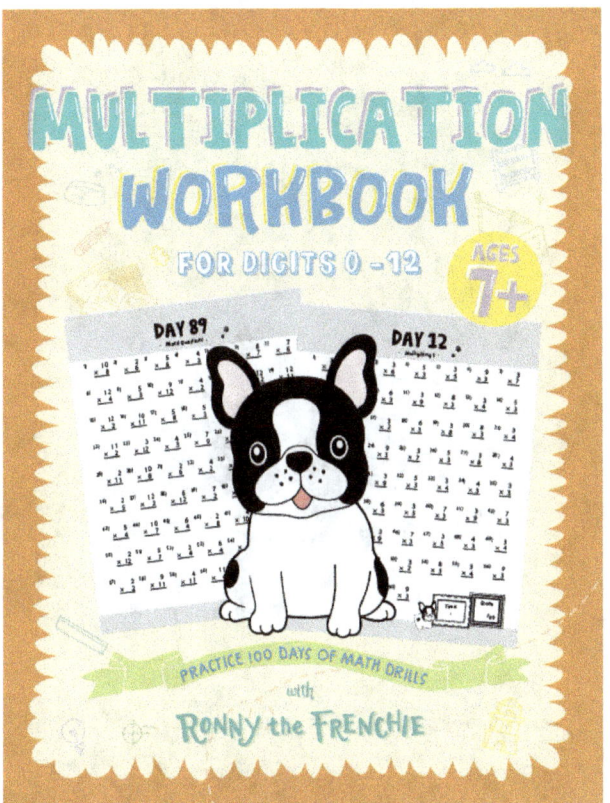

Multiplication Workbook for Digits 0-12

Did you know that there are many different names for the number 0? These include zero, nought, nil, zilch, and zip. There are 6 different names for the number 0! "Why did you only list 5 numbers?" you may ask. Well, to find out the historical sixth number that 0 used to be called, you'll have to solve the equations to get to the facts and find out!

Join Ronny and watch your kids become great at multiplication as he teaches them awesome and amazing math facts along the way.

Addition & Subtration Workbook For Double, Triple, & Multi-Digit

Have you heard of Carl Friedrich Gauss? How about the ancient counting tools? Ronny the Frenchie will explore these facts and more in a journey through his fact filled, knowledge-devouring brain. Find out how Gauss added together all the numbers from 1 to 100 so quickly and where negative numbers first appeared!

Join Ronny and watch your kids become great at addition and subtraction as he teaches them awesome and amazing math facts along the way.

Long Division Workbook: Learn to Divide Double, Triple, & Multi-Digit

Learn more about long division with Ronny the Frenchie, your favorite math friend! He's back with more amazing and astounding facts to share with you. Progress your learning of division and have fun at the same time!

What is the smallest number divisible by 1, 2, 3, 4, 5, 6, 7, 8, 9, and 10? Ronny the Frenchie will let you know how he discovered this fact, but you'll have to complete some pages in the workbook before you'll find out!

www.ingramcontent.com/pod-product-compliance
Lightning Source LLC
Chambersburg PA
CBHW072106110526
44590CB00018B/3336